Special Report:
THE AFTER-ACTION CRITIQUE:
Training Through Lessons Learned
April 2008

Reported by: Joseph Ockershausen

This is Report 159 of Investigation and Analysis of Major Fire Incidents and USFA's Technical Report Series Project conducted by Tri-Data, a Division of System Planning Corporation under Contract (GS-10-F0350M/HSFEEM-05-A-0363) to the DHS/USFA, and is available from the USFA Web page at http://www.usfa.dhs.gov

 FEMA

Department of Homeland Security
United States Fire Administration
National Fire Data Center

Department of Homeland Security
U.S. Fire Administration
Major Fire Investigations Program

The U.S. Fire Administration (USFA) develops reports on selected major fires throughout the country. The fires usually involve multiple deaths or a large loss of property, but the primary criterion for deciding to write a report is whether it will result in significant "lessons learned." In some cases these lessons bring to light new knowledge about fire–the effects of building construction or contents, human behavior in fire, and so forth. In other cases, the lessons are not new, but are serious enough to highlight once again because of another fire tragedy.

Under this project, USFA also develops special reports addressing a variety of issues that affect the fire service, such as homeland security and disaster preparedness, new technologies, training, fireground tactics, and firefighter safety and health.

The reports are sent to fire magazines and are distributed at national and regional fire meetings. The reports are available on request from USFA. Announcements of their availability are published widely in fire journals and newsletters.

This body of work provides detailed information on the nature of the fire problem and the many types of service provided by fire departments. The information informs policymakers, who must decide on allocations of resources between fire and other pressing problems, and personnel within the fire service, to improve codes and code enforcement, training, public fire education, building technology, and other related areas.

For reports on major fires and other critical incidents to which fire departments respond, USFA, which has no regulatory authority, sends an experienced fire investigator into a community only after having conferred with the local fire authorities to ensure that USFA's assistance and presence would support, not interfere with, any local review of the incident. The intent is to arrive after the dust settles so that a complete and objective review of all the important aspects of the incident can be made. Local authorities review USFA's report while it is in draft form. The USFA investigator or team is available to local authorities should they wish to request technical assistance for their own investigation.

For additional copies of this report write to the United States Fire Administration, 16825 South Seton Avenue, Emmitsburg, Maryland 21727 or visit our Web site: http://www.usfa.dhs.gov

U.S. Fire Administration
Mission Statement

We provide National leadership to foster a solid foundation for local fire and emergency services for prevention, preparedness and response.

TABLE OF CONTENTS

INTRODUCTION

The fire service is always seeking ways to improve its operations. At training classes and seminars, fire service members seek out insight into tactics and discuss new technologies for their applicability to other jurisdictions. Lessons also are learned from each response to an emergency incident. Unfortunately, many of those experiences and lessons are limited to those who were involved directly in responding. Unless feedback on incident response and command is shared with other personnel in the fire department, a valuable learning opportunity can be lost. The fire service has a duty to its members and the community it serves to evaluate problematic incidents, as well as those that go extremely well, and communicate the findings (including the lessons learned), to all relevant emergency personnel. An effective way to accomplish this is through a postincident critique.

The critique can be a powerful tool for effecting change. A noted fire service expert[1] observed that, "the post incident critique allows emergency responders to get a clear idea of the effects of their actions on the outcome of the operation. By comparing the expected outcome to the actual consequences, the fire department can make personal as well as organizational adjustments. And by assessing what worked, and what did not, improvements can be made."

A critique is a fact-finding exercise and a chance to relate and record pieces of information that collectively form a picture of the event and how personnel responded from both a command (tactical) and line (operational) standpoint. It is a tool to assess firefighting, rescue, and training effectiveness, and should include tactical plans and command decisions accompanied by how well they were followed. Lessons learned from the experience should be used constructively to correct deficiencies and influence training and education. Changes made to the department's plans and procedures typically occur per the outcome of incident critiques. Management must be willing to act upon the lessons learned and correct the problems as quickly as possible; otherwise, subordinate personnel will think the critique process is a waste of time, and future critiques could suffer accordingly.

The term "critique" may carry a negative connotation for some personnel in the fire service. Critiquing an incident may be perceived as a way to assign blame for mistakes that were made. The postincident critique, (or perhaps a less threatening term such as debriefing, after-action review, or postincident analysis (PIA)) should be viewed as a constructive way to obtain helpful feedback and positive suggestions. The process should be considered an important tool for improving firefighter safety and health, as well as a means for ensuring that the public is receiving quality services.

THE NEED FOR CRITIQUES

The number of fires in the U.S. that are responded to by the fire service continues to decline, even as the population increases.[2] While that is certainly a positive trend, fewer fires lead to less actual firefighting experience for firefighters, and sometimes reduction in the number of uniformed personnel in the fire departments.

---

[1] Carter, Harry R. "Post-Operation Critiques." *Firehouse*. Feb. 2001.

[2] *Fire in the United States 1992–2001*, Thirteenth Edition, U.S. Fire Administration/National Fire Data Center.

Historically, reductions in force played out as "last hired, first fired," so people with seniority generally were safe. However, because of the rising cost of pension plans and other retirement benefits that have skyrocketed, and because senior personnel make higher wages than newer firefighters, many jurisdictions now offer senior personnel buyouts and early retirement incentives to maximize cost savings. The resulting situation is that as senior personnel leave the fire service, their experience goes with them. Most of these individuals have 20 plus years of experience at both the command and operations level. Even if these positions are replaced with newly-trained personnel, those individuals will have less opportunity to hone their skills through actual firefighting or through personal experiences passed on by older firefighters.

The postincident critique is one way to bridge the experience gap by conducting routine after-action reviews. The department compiles and catalogs the problems encountered and the successful actions taken, thus creating a repository of information encompassing a wide variety of subject matter. The fire department can use the lessons learned to improve plans and procedures. Personnel quickly learn that their actions or inactions at an incident might be reviewed as a means of correcting problems, or, conversely, indicating that training was effective and the firefighters correctly applied the skills they learned. If problems show up, personnel at all levels will discern what went wrong and how to improve the outcome the next time. The following are examples of the inherent value of critiques and what they can accomplish:

- provide emergency service personnel with a clear indication of the impact their actions had on the general outcome of an incident;
- used to analyze and compare how different applied strategies and tactics affect the outcome of incidents;
- identify trends and patterns in errors during emergency operations so that immediate action can be taken to prevent them from reoccurring;
- identify positive outcomes that reflect proper attention to procedures, good decisionmaking, leadership skills, and so forth;
- serve as a catalyst for revising flawed tactical plans and Standard Operating Procedures (SOPs);
- used as a test bed where alternative tactics and evolutions are attempted, and to study their effect on the outcome of the incident;
- help identify additional or remedial training for personnel;
- used as technical reference material and cataloged for retrieval and examination during any similar future incidents;
- disseminate critical lessons learned during an incident to personnel throughout the fire department;
- identify fire prevention and code enforcement deficiencies;
- determine the need to install fire detection and suppression systems;
- identify illegal and required modifications to structures;
- identify structural and fire protection system failures; and
- identify built environment and operational challenges that contribute to civilian and firefighter injuries and fatalities.

ESTABLISHING POLICIES AND PROCEDURES

The key to successful critiques is having and enforcing a written policy that establishes a systematic and standardized approach for conducting them and clearly explains the purpose and objectives of evaluating the department's responses to incidents. The policy should lay out the process and define which types of incidents will undergo a critique: informal or formal. By establishing a critique policy, the fire department can ensure that every after-action review is conducted in a consistent manner and achieves the goals intended.

Some fire departments find it helpful simply to adopt another department's critique policy, and then modify it to fit their needs. The World Wide Web is another good source for finding excellent critique policies; many fire departments post their SOPs and other procedures on the Web. While conducting research for this report, a simple Web search turned up many comprehensive critique procedures.

Fire department leadership must create an environment that promotes trust and encourages personnel to participate openly and honestly in the critique process without fear of personal attacks or official retribution. Otherwise, some personnel may withhold valuable information that could be beneficial in determining how to improve future operations, and enhance the safety of personnel.

It is important when developing a critique policy that it be a collaborative effort. One approach used by many fire departments is to appoint a committee comprised of both line and staff personnel when drafting the critique policy, which helps dispel much of the cynicism associated with critiques. In departments with labor organizations their representatives should be encouraged to participate.

FREQUENCY OF CRITIQUES

A critique should not be held merely for the sake of doing one; there should be a good reason. The decision whether to conduct a critique or not should be made as soon as possible following the incident. The best time to conduct an informal critique is immediately following an incident when emergency personnel and units are still on the scene, and while the information is still fresh, but this may not always be possible because some departments have limited resources that must be made ready for the next emergency, or they may have a high call volume. As will be shown later, in such situations the critique can be conducted after the company has returned to quarters.

Since truly major incidents occur infrequently, the majority, if not all of them, should be critiqued. Small-scale incidents occur more frequently, so it is harder to decide how many, or which incidents should be critiqued without going overboard. There are several approaches that fire departments can take for selecting which incidents to critique. One method is to narrow the selection criteria by predesignating specific types of incidents that would automatically trigger a critique. Such incidents might include

- multiple alarm fires;
- hazardous material incidents;
- fires with injuries or a fatality;
- fires that exceed a predetermined dollar loss;
- technical rescues;
- incidents with unusual circumstances or unexpected development;
- fires in high-risk buildings; and
- incidents in structures where fire protection features influenced event outcomes.

A more conservative approach would be to grant Command Officers discretionary authority to select which incidents to be critiqued. The Incident Commander (IC) is likely to be more aware of the strengths and weaknesses of the personnel in his/her command, and therefore is in a better position to gauge what type of incidents personnel might most need to evaluate. No matter what strategy is selected, it is important that the selection process be incorporated into written critique procedures.

Critiques of small-scale incidents can be as beneficial in detecting trends and repeated errors made during emergency operations as large-scale events. If mistakes are made during small-scale incidents, it can be assumed that similar ones are being made during large, complex incidents. It could be difficult to detect such problems without having a series of critiques from which to compare emergency operations. The success of future operations depends on the early detection and correction of such problems.[3]

TYPES OF CRITIQUES

A postincident critique can be as simple as crewmembers sitting around a table in the station discussing their actions after returning from an incident; or it can be a comprehensive structured process that requires advance preparation and coordination.

The research for this report revealed several critique formats currently in use by various fire departments: individual or personal critiques, informal group critiques, semiformal and formal critiques. The most common methods used are informal and formal reviews. The other critique formats are basically variations of the informal and formal critiques. In most cases, the level of critique is dictated by the type, size, and complexity of the incident. Others factors play a role as well, for example, a fire where personnel had a close call with injury or a serious case involving injuries or deaths or where communication problems complicated response.

THE INFORMAL CRITIQUE

The informal critique is normally used at the company level (though multiple companies might be involved as well) and typically covers a review of how well specific tactics worked and what changes might induce better results. It is conducted on a case-by-case basis mostly for training purposes, and for the overall improvement of fire department operations. The critique can be initiated by either the Company Officer (CO) or Command Officer immediately following an incident as determined by department-based criteria. Ideally the assessment is done while emergency crews and apparatus are still on the scene, and while hoselines are still deployed. This provides optimal conditions for retracing the actions of crews and for analyzing any tactical or operational problems that emergency crews may have encountered during the incident. This is also an excellent opportunity to suggest alternative tactics and fireground evolutions and assess their potential impact on the outcome of the incident.

It may not always be possible or desirable to conduct a critique at the fire scene. Inclement weather, the time of day, and excessive call volume might prohibit units from immediately evaluating their response. In these situations, the CO can conduct an informal critique after returning to quarters.

During the informal critique, the CO serves as the moderator to ensure the discussion stays on track. The critique should begin by the officer reviewing the company's assigned objectives during the incident. Each crew member should be given the opportunity to explain his/her assigned tasks, any problems encountered, and actions taken during the incident.

---

[3] Morris, Gary. "The Incident Critique." JEMS. Sept. 1987, p. 38.

The officer should take the time to recognize exceptional performances by personnel. Everyone deserves recognition for a job well done, and positive reinforcement goes a long way to improve an individual's confidence and morale. This is especially important if the firefighter also made a mistake that had an adverse impact on operations. If an individual firefighter failed to perform as well as expected during an incident, admonishing him or her in front of the rest of the crew will serve no useful purpose. It is embarrassing, will undermine the self-confidence of the individual, and affect company morale. Any adverse performance-related discussion with employees should be conducted in private, but they should take place, especially since mistakes can place other firefighters at greater risk. The rest of the crew needs to know that all problems will be addressed, even if some are handled behind closed doors.

If operations failed to go as planned, the problems need to be identified and corrected. The officer has many options when considering corrective action. These may include simply reviewing the department's SOPs with personnel, arranging for remedial individual or company training, or increasing company-level training. Remember, the informal critique is a **learning exercise**; the point is to acknowledge success or correct and prevent similar mistakes from being made in the future, not to publicly embarrass those responsible for making them.

It is important that officers include themselves in this exercise by explaining their responsibilities and actions during the incident, and admitting to any tactical mistakes they may have made. Once the firefighters know the officer is willing to acknowledge a less-than-perfect performance, they will more readily admit their own or identify mistakes by others. Seeing the incident from the officer's perspective and understanding the methodology behind the decisions gives the crewmembers a better understanding of what is expected of them during an incident. To maximize the learning experience, sufficient time should be allocated to complete the review. The objective is for the entire crew to get a clear picture of the impact that its actions had on the outcome of the incident, and what it could have done better to improve the outcome.

An informal format is sometimes used by COs and Command Officers to meet one-on-one with newly-appointed firefighters and officers to discuss their performance. The review of actions and decisions lets the officer praise subordinates for jobs well done, and provide new firefighters and officers detailed guidance for improving performance, which is beneficial to their professional development.

Normally an informal critique is not documented. However, if a unique situation is encountered that might affect emergency operations in either a positive or a negative way, the CO should forward the information gleaned to the next level in the chain of command for further consideration.

THE FORMAL CRITIQUE

The formal critique is a detailed review and analysis of large-scale and other complex or tactically-challenging operations. Examples include natural disasters, terrorist events, mass casualty incidents, major fires, fires involving multiple injuries or deaths, firefighter line-of-duty deaths, building collapse, large-scale hazardous material incidents, transportation disasters, and major wildland fires. These types of incidents normally involve a large-scale response and, often, assistance from other outside agencies. Many fire departments never experience a major incident. However, when one does occur, the department's ability to respond effectively will quickly be put to the test.

The formal critique is used to reconstruct an incident to determine if the department had an appropriate tactical plan and procedures and if they were followed, as well as how effective they were in mitigating the incident. Every aspect of the incident is carefully reviewed and analyzed to determine what went well, what could be improved, and why. The results of such analysis can suggest changes to a department's plans and procedures that may be necessary.

The next step is an all-hands critique meeting which should be scheduled as soon as possible and practical after the initial incident. All relevant personnel should be notified of the date, time, and place where the critique is to be conducted. Scheduling the critique may be difficult, especially when shift work is involved. The critique should be held on a day that the same shift is scheduled to work, so that the personnel who were involved in the incident will be available. If possible, the critique should be scheduled within a few days of the original incident; any longer a delay and personnel may forget important details concerning the incident. The session should be for responders, not the press or political office holders. However, invitations also should be extended to fire department personnel and mutual-aid companies from neighboring jurisdictions as well as any other agencies that may have been involved in the incident. In some cases, it may not be possible for agencies to send everyone involved in the incident to the critique. In this case, at least the agencies' personnel who served in key positions during operations should attend.

If possible, the critique meeting should be held at a central location in an effort to reduce travel distances, and to permit as many people as possible to attend. The facility should be large enough to accommodate everyone comfortably, and have the necessary audiovisual equipment and other necessities on hand.

The meeting should last long enough to accomplish its intended goal, but not beyond a reasonable point. Two to three hours of focused discussion and review, with a short break in the middle, should generate a lot of good information.

Following the critique, the fire department should brief appropriate local government officials as to its preliminary findings. Local officials often are besieged with inquiries from citizens and the press after a major incident. Local government officials need accurate information so they can be responsive and provide accurate information to the public.

Once a determination has been made to critique an incident, a critique officer should be appointed. For formal critiques, the critique officer should be a chief officer who is in a position to promote change if a critique sheds light on needed improvements. If not, little will be learned from the critique. Mistakes made during the incident will go unresolved, and probably will continue to be a problem during future operations. The critique officer is the glue that holds the critique process together. The critique officer should be responsible for the overall preparation and coordination of the critique, and ensure that the information and equipment necessary to conduct the critique is assembled in a timely manner. Emotions can run high during a critique, and things can get out of hand if control is not maintained. The critique officer should maintain firm control over the exercise by establishing and conveying the ground rules, and by encouraging personnel to participate openly in the process. This individual plays a critical role in the critique process and his/her selection should not be taken lightly.

The critique officer should have excellent organization, communication, and people skills. The IC or Safety Officer might be the logical choice to assume this position because of his/her intimate knowledge of the tactical and operational plan, and any problems that may have been encountered during operations. However, this may not be the best choice, especially if things did not go well during the incident. Because of his/her close involvement in the incident, an IC may find it difficult to be objective or to avoid being defensive. In such situations an uninvolved officer might be a preferable choice to lead the review.

INCIDENT DATA COLLECTION

One of the first things the critique officer should do is develop a checklist of information that should be collected and reviewed. As much information about the incident as possible should be obtained from multiple sources.

An important element of operational critiques is that questions are consistent within each type or level of personnel. For example, suppression personnel would be asked the same questions; emergency medical services (EMS) members would have a set of questions, and so forth. In this way, the opportunity to identify trends and repeated problems during emergency operations will be enhanced. Even though every incident is different, command and control of the incident should generally adhere to the department's tactical plans and SOPs. Therefore, the critique should determine if the department's plans and procedures were followed, and if so, how effective they were.[4]

One way to maintain consistency is to use a postincident questionnaire. A growing number of fire departments now use some form of questionnaire to collect information after an incident. Questionnaires can be used for any level critique; however they always should be used during formal critiques of major and complex incidents in which the department had significant resources committed. Fire departments that use the National Fire Incident Reporting System (NFIRS) report can obtain much of the basic information regarding the incident to be critiqued from the various NFIRS report modules.

As part of the USFA incident management curriculum, comparative PIA questionnaires have been developed that can be used by COs and Command Officers, and the critique facilitator (critique officer). Using a standard questionnaire is a fast and efficient way to collect information for the critique. A copy of the comparative PIA questionnaires can be found in the Appendix.

Using the example questionnaire developed by the USFA, some departments may wish to add a section on who would be responsible to address needed changes and when and how the proposed improvements will be implemented. The information gleaned from questionnaires can provide the critique officer a good understanding of what occurred during the incident, and the problems encountered during operations. A list of discussion points can be developed then to ensure that all relevant issues and concerns regarding the incident are addressed adequately during the critique. The questions also will help keep personnel engaged in the critique process and help keep the critique on track.

Once a determination has been made to review how the department responded to an incident, questionnaires should be sent within a couple of days of the event to all levels of the fire department that participated in the incident. It is important that a firm deadline be established for completing and returning the questionnaires. Officers need to monitor compliance. Responses are needed quickly so that there is ample opportunity for the critique officer to compile and analyze the results.

---

[4] Morris, op. cit.

The response results will form the basis for both the all-hands critique meeting that comes shortly thereafter, and for the written after-action report.

The material developed by the USFA represents only one source of information about the incident. The following table contains examples of the types of reports, data, and questions that should be covered for a formal critique. Although a questionnaire can be used for either an informal or a formal review, they should always be used for the latter. These examples of incident information and questions were compiled from various sources including the Phoenix, Arizona, and Charlottesville, Virginia, Fire Departments; Sonoma County, California's, EMS agency; *Firehouse* and *JEMS*; International Society of Fire Service Instructor (ISFSI) instructor critique guide; and other related sources. The data collection array has been consolidated per the organizational functions within a fire department, which includes fire suppression, fire prevention, and special and support services.

Information Needed for Blueprint of Major Incident Critiques

Fire Suppression

IC	Fire Suppression Units
• Date and time of the incident.	• What conditions confronted personnel upon their arrival?
• Incident location.	• Describe apparatus deployment.
• Weather conditions at the time of the incident.	• Provide sector assignments and assigned objectives.
• Type of occupancy (fixed property use).	• Was the initial water supply adequate? If not, what was done to provide adequate supply?
• Topography of the incident scene.	• What size supply lines were deployed and where?
• Describe conditions upon your arrival; specifically, did the conditions warrant offensive or defensive tactics?	• Describe the position and size of attack line(s).
• Identify all problems encountered and actions initiated to overcome them, if possible.	• Was the initial attack line adequate?
• Did transfers of Command go smoothly?	• Were any operational problems encountered during the incident?
• Describe actions initiated by first-arriving units. Were they appropriate?	• Describe the events/actions the crew initiated to obtain assigned objective.
• Provide site drawings of incident layout, apparatus positioning, attack line placement, Rapid Intervention Teams (RITs), accountability locations, Staging, firefighter rehab, etc.	• Describe all events/actions that hindered accomplishing assigned objectives.
• Provide overview of responsibilities and activities assigned to each section.	• Were any safety problems encountered during the incident?
• Describe rescue problems encountered, and actions taken to overcome them.	• Did the crew experience any equipment failures during operations?
• Describe ventilation problems encountered, if any, and the steps taken to overcome them.	• What could be done differently next time to improve operations?
• Was the initial assignment adequate to handle the incident, and if not, what additional units were requested and why?	• What changes are recommended to existing plans and procedures, or training as a result of the incident?
• Describe how many and where RITs were deployed.	
• Describe any exposure problems, and the steps taken to protect them and so forth.	
• Describe all safety-related concerns.	
• Develop a sequential incident organization chart detailing Command and Group/Division assignments.	
• What could have been done differently to change the outcome of the incident?	

Information Needed for Blueprint of Major Incident Critiques (cont'd)

Support Services

Dispatch/Fireground Communications

- Provide audiotapes of the phone reports, dispatch, and tactical communications to the point where the fire was reported under control.

- Did the IC make routine incident updates and at what intervals?

- Provide computer-aided design (CAD) printout of the incident timeline.

- Identify any communication problems encountered during the incident, and the steps taken to resolve them.

Logistics/Fleet Maintenance

- Identify all support assets committed to the incident.

- Identify all responsibilities assumed by the support asset and their effectiveness (i.e., firefighter rehab, logistical support, financial, fleet maintenance, etc.).

- Provide a summery description of equipment or apparatus performance, repairs, refueling, etc.

- Identify all problems encountered during the incident.

Public Information Officer (PIO)

- Obtain copies of raw news media video, photographs, etc.

- Provide overhead projector and white board for illustrating the incident.

- Produce transparencies of incident depicting apparatus positions and attack line deployments and sector assignments.

- Provide site photography, including aerial shots, if possible.

- Obtain copies of video taken by civilians and/or residents of the community.

Fire Prevention

Code Enforcement

- Provide a legal description of the structure, including the number of floors, basements, type of construction, type of roof structure, mechanical systems, number of and locations of exits.

- Describe the type of fire detection and suppression systems in the building, and whether they functioned properly.

- Were the building fire suppression systems effective in containing and/or extinguishing the fire?

- Describe the effects the fire had on the structure.

- Did the structure suffer either a partial or a total collapse?

- Provide a historical overview of all building fire inspections, code violations, use of occupancy certificates that have been issued, and a description of all modifications made to the building.

- Ascertain whether the structure met current building and fire codes and identify all outstanding code requirements not met.

Fire Investigations

- Provide a description of the origin and cause of the fire, including type of ignition (i.e., accidental electrical, suspicious, flammable liquids, etc.).

- Describe the factors that influenced fire spread (i.e., were accelerants used, attack lines placed improperly, ventilation techniques employed improperly).

- Did the fire suppression actions compromise the building's structural stability?

- At any point during the fire did the structure pose hazards to firefighters? If so, where?

- Provide a value and loss assessment of the structure.

- Describe the type of fire detection and suppression systems the building was equipped with, and whether they functioned properly.

- Provide a list of all structural design features, protective systems, and other building components that were not provided, that would have reduced the spread of fire, fatalities, and injuries, or would have reduced property loss.

Fire Prevention (cont'd)

	• Is there a link between this fire and other fires in the area?
	• If possible, obtain photographs, slides, or video taken by investigators that may be helpful to the critique process.

Special Services

EMS

- Provide a summary of basic life support (BLS) and advanced life support (ALS) units dispatched on the initial response. Were additional EMS required? When were they requested?

- Were EMS supervisory personnel on the emergency scene?

- Were medivac helicopters used to transport patients?

- Were EMS communications adequate?

- Were multiple medical control communication points established and with what medical facilities?

- Provide a summary on patient distribution by hospital, number of patients to each hospital, triage category at site, hospital priorities, etc.

- Provide followup summary report on patient injuries, expected outcomes, etc.

- Describe what went well, and what could have gone better.

- Was a critical incident debriefing team used?

Hazardous Materials (Hazmat) Incident

- Provide a summary of Hazmat response and evaluation by section officer.

- Provide a summary of Hazmat mitigation efforts and resources deployed.

- Provide description of hazardous materials present and level of management and procedures.

- Provide description of exposure hazards to firefighters and suggested incident.

- Provide copies of Department of Transportation (DOT) guidelines and other related literature and reference sources used.

- Provide a summary of cleanup operations.

- Provide lessons learned from any in-house critiques of Hazmat incident activity.

PROCEDURES FOR THE FORMAL CRITIQUE MEETING

It would be a good idea to develop an agenda when conducting a formal critique meeting. An agenda will help to focus the discussion and ensure that the important issues are addressed. Remember, time is critical during a critique, and the critique officer can ill afford to waste it with discussions that evolve as freelance speeches. Documenting the incident scene and the actions of emergency crews is an essential part of the critique process. Equipment such as an overhead projector and easel pads are very helpful when configuring the incident scene. Emergency personnel can use these devices to illustrate the positioning of apparatus, sector assignments, and deployment of attack lines, water supply points, exposures, and hazard zones in relation to the incident scene. Other materials such as writing supplies should be provided for note-taking purposes and jotting down questions to be asked later. Audio equipment should be available during the critique to permit personnel to listen to the 9-1-1 call and incident communications, if appropriate. When using audiovisual equipment, ensure that it works properly before the critique, and that replacement bulbs, batteries, and transparencies are on hand should they be needed.

Wallboards, wet boards, or similar devices should not be used for recording information about the incident because they are easily erased, and crucial information could be lost. Following the critique, steps should be taken to preserve the charts and transparencies documenting the incident scene. From this information, scale drawings depicting the incident scene should be made and incorporated into the final report.

The critique officer should appoint someone to take notes during the critique in preparation for the final report. Special emphasis should be placed on the problems encountered, solutions, and failed remedies. To improve retention and accuracy, the critique meeting could be videotaped. Recording the session will ensure that the information and issues discussed and the lessons learned during the meeting are preserved for training purposes and use by future emergency personnel. Some jurisdictions have local cable channels dedicated specifically for use by public safety. This is an excellent communications medium to deliver videotaped critiques to everyone throughout the fire department. Videotapes of major incidents can be cataloged and archived for future training purposes. For departments that lack this capability, the videotape can be duplicated cheaply and distributed with a copy of the after-action report to the rest of the personnel in the department.

The following are basic topics that should be addressed during the formal critique:

Introduction–The critique officer should begin with a short introduction. The officer should explain the purpose and the objective of the critique, and stress the importance of the critique process and how it will benefit the fire department and public safety in the future. The personnel should be encouraged to participate in the critique, and ensured that the purpose of the critique is to identify successes and areas where improvements need to be made, not to assign blame. Participants should arrive having completed the questionnaire relevant to their role in the response, and the questionnaires should be collected at the sign-in desk.

Ground rules–The critique officer needs to establish the ground rules at the beginning. The critique is intended to be a constructive process, and those in attendance should be reminded to conduct themselves in a professional manner during the critique. Conduct such as finger pointing or derogatory comments directed at individuals or crews is counterproductive and should not be tolerated. The critique should have a time limit. However, meaningful dialogue regarding the incident should be permitted by the critique officer as long as the discussion does not get off the topic or become a sounding board for grumbling and complaints. The critique officer should request that all questions be held until the overview of the incident is finished, and the discussion session begins. This should help keep the critique moving along and avoid wasting time belaboring unimportant issues. A time limit should be established for each speaker and ensure that no one individual dominates the critique.

Overview of incident–The critique officer should provide a brief overview of the incident that includes the type of the incident, date, time, type of structure, occupant load, location, and weather conditions at the time of the incident. If applicable, the tactical preplan should be reviewed to familiarize personnel with the structure or conditions (hurricane, mass-casualty event, etc.) that are the focus of the critique. The critique officer should describe the situation as it was reported to the Dispatch Center, and provide a list of all the equipment dispatched on the incident including any additional alarms or special request apparatus and mutual aid received.

Incident review–Every aspect of the incident should be reviewed, including decisions made by the IC, Division officers, and line companies. The critique should begin by having the officer on the first-arriving unit to the scene illustrate the incident scene: where they positioned their apparatus, what conditions confronted them, and what their initial actions were. This process should be repeated in the sequence in which emergency units arrived on the scene of the incident. The final Incident Command System (ICS) structure should be displayed on an overhead projector, poster, or handout for all participants to reference during the discussions. The fastest way to gather information is by asking open-ended questions which require a direct response, such as, what did the crew observe upon arrival? What information did they receive? What problems were encountered? Were the problems reported, and to whom? How were the problems overcome? The same process should be followed for the Command element of the operation.

Discussion points–The critique should conclude with an open discussion about the participants' observations, specifically, how fire controlled objectives, and how strategic and tactical efforts unfolded during the operation. What worked well? Where is improvement needed? Were the department's tactical plans and procedures applied, and if so how effective were they? Do they need revision or modification? What remedial or additional training should be scheduled? These are the basic questions that should be addressed. The final report should answer these and any other questions or concerns identified during the critique.

Performance recognition–Exceptional performances by personnel should be acknowledged during the critique. Everyone enjoys a pat on the back, and recognition for a job well done will encourage others to do well too. A poor performance should not be discussed during the critique; it should be done privately, as in the case of informal critiques.

Report–The most important part of the critique process is the report. This document summarizes the entire incident and provides recommendations for correcting problems based on the lessons learned during the critique. It also can serve as a blueprint for additional training and the development of better plans and procedures.

The report should begin with basic information about the incident such as a the type of incident, the date, time, and place where it occurred, weather conditions, number and type of apparatus dispatched, and information on any injuries and fatalities. This should be followed by a narrative section that describes the key aspects of the incident in greater detail, including the conditions confronting emergency crews, the problems encountered by crews, life safety issues, and strategic and tactical operations including diagrams of the incident showing apparatus deployment and group or division assignments. Each area of assignment during the incident should be addressed in the narrative section. This component may vary depending on the size of the department and complexity of the incident. Assignment areas that should be considered include Incident Command, objectives, strategy and tactics, dispatch and communications, emergency medical, safety, apparatus and other resources, and mutual-aid assistance.

The next component of the report should be the lessons learned. Nearly every critique will reveal problems that need to be addressed to improve future operations. Problems experienced during operations are generally attributed to ineffective plans or procedures, or improper actions by personnel. Problems related to the department's SOPs may require the department to revise existing SOPs or create new ones. Performance-related issues are generally training related. Personnel are either unfamiliar with relevant policies and procedures, or have little or no practical experience in a given area. This situation often can be corrected through additional training and education. The department leadership routinely should compare critique findings to detect trends or recurring problems that may signal deeper problems that may need to be addressed.[5] The final component of the process should be a management approved action plan to adopt the issues addressed in the lessons learned portion of the critique.

The following case studies are examples of successful critiques. They illustrate how a critique can have a positive effect in initiating change and improvements, which enhance tactical and operational performance toward the goals of assuring public safety.

---

[5] Morris, op.cit.

CASE STUDIES

Phoenix Fire Department Firefighter Fatality

In 2001, the Phoenix, Arizona, Fire Department suffered a firefighter fatality in a five-alarm fire that trapped one and injured three firefighters. Following the fire, and for the next year, dozens of fire department committees analyzed all aspects of the fire including building codes, personnel, equipment, SOPs, technology, and training. The critique led to numerous recommendations and changes in the City's building codes, and the Phoenix Fire Department. The following is a brief overview of the recommendation in the final after-action report.[6]

The Bert Tarver Sprinkler Ordinance (named after the deceased firefighter)

The final after-action report recommended revisions of the city's sprinkler code, which expanded the requirement for automatic sprinkler protection in new and existing residential and other occupancies. Many of the city's existing residential and other structures that have undergone renovations and meet the new square footage provisions as of June 2002 will be required to retrofit with sprinklers under the new sprinkler provisions. Expanding sprinkler protection will have a positive impact on the city's fire problem and improve firefighter safety. Sprinkler systems have been extremely effective in reducing property damage and the loss of life due to fire. According to the National Fire Protection Association (NFPA) as few as two properly functioning sprinkler heads will contain and/or extinguish a fire in its incipient stages.

Personnel

The report calls for additional Command and operational positions within the department to improve incident management and safety capabilities. Additional personnel would be assigned to selected truck companies; heavy rescue companies would be dispatched to working fires to bolster staffing. Enhancements to other key positions include additional Safety Officers, Training Officers, Tactical Preplanning Officer, and additional personnel to staff the mobile Command Post (CP), and provide an additional full-time position at the departments alarm room (Dispatch Center). Adding additional Command and tactical personnel will help improve Command and Control, safety, and communications during emergency operations.

Standard Operating Procedures

The report concluded that although effective SOPs were in place at the time of the fire, Command and field personnel lacked sufficient task-level training on the procedures to react effectively to the incident. The final report recommends that the following SOPs undergo various levels of revision: accountability, Rapid Intervention Crews (RICs), rescue lost firefighter command responsibility, safety sector, lost or trapped firefighter basic self-survival, and "may day" communications. It was recommended that these SOPs be combined into a new training manual and training lesson plan for firefighter safety and survival, schedule regular training not to exceed 2 years, and put formal training in place regarding new management procedures for firefighter safety and survival. The new emphasis on training will enhance firefighter safety and survival through better Command and Control and firefighter survival skills.

---

[6] Phoenix Fire Department's Final Report on the Southwest Supermarket Fire.

Technology

The fire department has been evaluating many new and innovative technology-based systems to enhance firefighter safety and survival. Some of the emerging technologies the fire department is currently following include firefighter tracking and locating devices. It has been recommended to equip each Battalion Chief vehicle with a downlink antenna to permit the IC to receive video from overhead news helicopters. Improvements should be made to the department's current radio system. Other technologies reviewed and evaluated include state-of-the-art thermal imaging cameras, and a self-contained breathing apparatus (SCBA) air capacity monitoring telemetry unit. These new technologies will enhance the IC Command and Control capabilities during incidents through more effective monitoring of the incident scene, and by enhanced capabilities in tracking and locating downed personnel.

Training

The joint recovery team recommended that new training modules be developed to address the following areas: air management, self-survival, fireground communications, RIC deployment and responsibilities, and search and rescue techniques. The training would consist of two phases; the first phase involves a walk-through and explanation of what occurred at the Southwest Supermarket fire. Phase two involves classroom followed by hands-on training to reinforce the classroom training. Firefighters will learn how to manage their air supply under different situations, and learn critical thinking and behavior skills when low or out of air. Firefighters also will learn survival skills as an individual and self-survival skills as a crew. Much like the air management scenario, personnel will be taught critical thinking and behavior techniques as they relate to self and crew survival during a fire.

Training also will be provided on how to improve personal communication skills by radio and face-to-face. Enhanced RIC training will be provided; the first phase is designed to provide a basic overview of the RIC operations on the fireground including task, tactical, and strategic levels of the operations. The second phase consists of a demonstration and hands-on practice session covering RIC search methods, packaging, and victim removal techniques. This training program coupled with the other firefighter survival training, new technologies, and sprinkler code modifications should dramatically improve firefighter safety and survival during emergency operations.

Strategic Level

Strategic level recommendations are more philosophical issues that can be addressed only by management, and eventually by the whole department. These issues include

- Develop a fire department philosophy on how and why firefighters act the way they do on the fireground.

- Incorporate the strengths of Cockpit Resource Management (CRM) techniques developed by the airline industry to manage routine aircraft emergencies into the fire department command team structure.

- Develop a way to identify and balance critical fireground factors against the risk to the community and firefighters. This will help the fire department to better understand what is critical and what is not.

- The fire department must develop a method for determining what the proper fireground strategy is before commencing operations.

- Once the critical fireground strategy is identified, the fire department must develop a common language the enables all levels on the fireground to share information.

- The fire department must adopt a philosophy in which small fires are handled the same as fires in complex structures. The same safety consideration concerns should apply in both situations.

- The fire department must strengthen its ability to save its own personnel through increasing its expectations and by enhancing the department's RIC capabilities.

West Warwick, Rhode Island, Station Nightclub Fire

On February 20, 2003, a tragic fire occurred at the Station Nightclub in West Warwick, Rhode Island, where 100 patrons died and approximately 200 were injured. The band that played before an estimated crowd of 450 people used pyrotechnics for special effects purposes during the opening performance. The pyrotechnics ignited highly-flammable polyurethane foam insulation lining the wall and ceiling of the platform area where the band was performing, resulting in the deadly fire.

Following the fire, an intensive investigation into the cause of the fire was conducted by the National Institute of Standards and Technology (NIST).[7] During the course of the investigation, Federal investigators examined all relevant model building and fire codes; other incidents with similar circumstances in places of public assembly; fire detection and suppression systems that were part of the structure; materials used in the construction and interior finish of the building; points of egress and process; and the fire department response to the incident. Federal investigators were able to develop new information, and confirmed published reports as to the initiating event, the reason for the rapid spread of the toxic smoke and fire, the difficulties encountered by patrons during egress, and the mass casualty situation confronted by the fire department.

The direct contributors to the large loss of life were found to be (1) the close proximity of the highly-flammable polyurethane foam to the pyrotechnics that started the deadly fire, (2) the Station Nightclub was not equipped with a sprinkler system, which resulted in the inability to suppress the fire during its early stages of growth, and (3) the inability of the exits to handle all of the occupants in the short time available for the fast-growing fire.

The investigation also concluded that strict adherence to the 2003 model codes available at the time of the fire could have helped prevent the tragedy. Changes made to the code subsequent to the fire made them more effective. Stronger enforcement of the code also would help strengthen public safety even further.

---

[7] Report of the National Institute of Standards and Technology (NIST) *Technical Investigation of the Station Nightclub Fire.*

Arlington County, Virginia–September 11th Attack on the Pentagon

Following the terrorist attacks of September 11, 2001, Arlington County, Virginia, produced a major report[8] that documented the full-scale response to the terrorist attack on the Pentagon, which lies within their jurisdiction. The department examined all of the emergency and postincident operations, and incorporated information from all agencies that participated in the response to the Pentagon. Arlington County received high marks for their strategic vision, tactical leadership, flexibility, breadth of capabilities, and technical competence that ultimately led to a successful response to the attack.

The final report also identified several problems that adversely affected emergency operations, and offered recommendations on how to remedy those situations. The report cited that the freelancing of fire units was a big problem and recommended that they be relieved immediately. However, this could be a logistical nightmare and enforcement problem for the IC. Once fire and rescue units have taken up a position on the incident scene there is little chance of returning units already committed. The fire departments' SOPs and mutual-aid agreements should clearly delineate the fire departments' callback policy regarding major incidents, and stress that freelancing will not be tolerated. Personnel reporting for duty from home should first report to their duty station, report in, and await their assignment.

The report questioned Arlington County's staff recall policies. Consequently, the county has implemented new staff-recall procedures and has developed an Internet-based group mail system and a personnel recall system. Another enhancement to the department's communication capabilities under consideration is issuing pagers to all firefighters and paramedics. Another problem stemmed from the lack of a fixed or mobile Command and Control facility. Using Federal Homeland Security grant funds, the department has since purchased a new mobile Command unit, and equipped all fire and EMS units with a mobile data terminal. The final report also suggested that the department review fire apparatus staffing levels to improve the speed of early search and rescue operations and provide for the safety of crews. The county has approved funding to hire additional personnel so that each unit has four personnel. Likewise, the technical team grew from 30 to 42 personnel, and the Hazmat team increased from 33 to 45 members.

The preceding examples show how critiques can lead to big changes in the fire department. However, not every incident involves a terrorist attack, death of a firefighter, or mass casualty incident. Not every fire department has a Federal grant or tax base sufficient to fund all the improvements that an after-action study may reveal to be advisable. Is there any point in evaluating operations in these cases? The unequivocal answer is "yes."

---

[8] Weiger, Pam. "Pentagon Report." *NFPA Journal*. Nov./Dec. 2002.

APPENDIX–POSTINCIDENT ANALYSIS

POSTINCIDENT ANALYSIS
Company Officer Input

I.   INCIDENT DATA

Alarm #:_____ Date: _____

Your Unit Number:_____ Dispatch Time: _____

Arrival Time:_____Alarm: ___1st___ 2nd___3rd___Other _____

Your Incident Supervisor:_____ICS Function:_____

_____

Emergency Type:_____ _____

Describe the Situation on Arrival:_____

II.   STRATEGY

What were the Strategies for the Incident? _____

_____

How Long did it Take to Achieve the Goals? _____

In what Sequence were the Strategies Achieved?_____

_____

How did you Determine What the Plan was?_____

Personal Observation:_____Briefing By: _____

III.   TACTICS

Describe the Tactical Assignments Given to you in Chronological Order: _____

_____

ICS Position that gave you the Assignments: _____

Coordination Required with? _____

Coordination Determined:_____ At Briefing_____During Operations

How did you Determine your Supervisor?

_____In the Directive   _____Observation

IV.   PROBLEMS ENCOUNTERED

Type:

_____ Coordination   _____ Staff Support

_____ Ineffective Equipment Use   _____ Communications

_____ Inadequate Personnel   _____ Equipment Failure

_____ Safety   _____ Too Many Personnel

_____ Other

Descriptive Account of Problems Checked:_____

_____

_____

_____

_____

_____

_____

_____

_____

Recommendations:_____

_____

_____

_____

_____

_____

_____

_____

_____

_____

V.   ICS ORGANIZATION

Draw the ICS organizational chart for your part of the operation. Start with your immediate supervisor and go up and down as far as you know.

POSTINCIDENT ANALYSIS
Subcommander Report

I.   INCIDENT DATA

Alarm #:_____ Date:_____ Name or Unit:_____

Division, Group, or Sector Assigned: _____

Dispatch Time:_____ Time Received Assignment: _____

Describe the Situation on Arrival:

_____

_____

II.   STRATEGY

What were the Strategies for the Incident? _____

_____

How did you Determine What the Plan was?

Personal Observation:_____Briefing By: _____

III.   ORDERS RECEIVED (STRATEGIC OR TACTICAL):

_____

_____

IV.   ORDERS GIVEN:

_____

_____

Engine _____

Engine _____

Engine _____

Engine _____

Truck _____

Truck _____

Other _____

Other _____

V.   PROBLEMS ENCOUNTERED

Type:

_____ Coordination   _____ Staff Support

_____ Ineffective Equipment Use   _____ Communications

_____ Inadequate Personnel   _____ Equipment Failure

_____ Safety   _____ Too Many Personnel

_____ Other

Descriptive Account of Problems Checked:_____

_____

_____

_____

_____

_____

_____

_____

_____

Recommendations:_____

_____

_____

_____

_____

_____

_____

_____

_____

_____

_____

VI.   ICS ORGANIZATION

Draw the ICS organizational chart for your part of the operation. Start with your immediate supervisor and go up and down as far as you know.

POSTINCIDENT ANALYSIS
Incident Commander Form

I. INCIDENT DATA

Alarm #:_____ Date:_____ Your Unit: _____

Dispatch Time:_____ Time Assumed IC:_____

Time Relieved:_____ Relieved By: _____

Describe the Situation on Arrival:_____

_____

_____

II. STRATEGY

Identify the Action Plan Strategy: _____

_____

Time to Achieve:_____

Describe the Tactical Sequence: _____

_____

_____

Changes Made to the Action Plan:_____

_____

_____

III. ORDERS GIVEN:

1st Alarm:

Engine _____

Engine _____

Engine _____

Engine _____

Truck _____

Truck _____

Other _____

Other _____

2nd Alarm:

Engine _____

Engine _____

Engine _____

Engine _____

Truck _____

Truck _____

Other _____

Other _____

IV.   PROBLEMS ENCOUNTERED

Type:

_____ Coordination   _____ Staff Support

_____ Ineffective Equipment Use   _____ Communications

_____ Inadequate Personnel   _____ Equipment Failure

_____ Safety   _____ Too Many Personnel

_____ Other

Descriptive Account of Problems Checked: _____

_____

_____

_____

_____

_____

_____

_____

Recommendations: _____

_____

_____

_____

_____

_____

_____

_____

_____

V.   ICS ORGANIZATION

Draw the ICS organizational chart for the incident after all units were working.

POSTINCIDENT ANALYSIS
Facilitator Narrative Description

I.   INCIDENT DATA

Alarm #:_____ Time:_____ Date: _____

Address:_____

Units Responding: _____

Time of Control: _____

Describe the Situation on Arrival:_____

_____

II.   STRATEGY

Strategy:_____

_____

Time to Achieve:

Describe the Tactical Sequence: _____

_____

_____

_____

III. ORGANIZATION

Describe Organizational Problems: _____

_____

_____

_____

IV.   PROBLEMS ENCOUNTERED

_____

_____

_____

V.   RECOMMENDATIONS AND CONCLUSIONS

_____

_____

_____

www.ingramcontent.com/pod-product-compliance
Lightning Source LLC
Chambersburg PA
CBHW082035190526
45165CB00020B/3164